Think! EARTH

By
John Cantellow

Text copyright © 2021 John Cantellow
All Rights Reserved

Contents

1 Introduction	**5**
2 Emergency	**7**
O2 Production	7
Freshwater	8
Food Production	8
Climate Change	11
Biodiversity Loss	16
Pollution	18
3 The Problem	**20**
4 Needs vs Wants	**25**
5 Think! EARTH	**28**
Deciding where to live	29
Choosing Accommodation	30
Choosing Home Energy	30
Deciding what to eat and drink	31
Deciding what to wear	31
Travel	31
Consumer Products and Services	32
Education, Training and Working	33
Pets	33
Charities	34
6 Conclusion	**36**
Bibliography	**38**

1 Introduction

With exponential population growth combined with exponential increases in consumption per person, we are accelerating towards the total destruction of the only planet that will support us.

There is a prophecy within the Christian Bible that life will go on as normal when the end suddenly comes. That is exactly what is happening now. We are living our lives as though we can continue this way indefinitely. The overwhelming evidence paints a much blacker picture. Whichever way we look at it, we are racing towards a precipice of self annihilation.

This is an emergency that demands urgent action if we humans are to survive. Each and every one of us must make substantial changes to the way we live, so that we live within our planet's capacity to sustain us. With every decision, every choice that we make, we must *Think*! EARTH.

It is worth pointing out at the outset that there are five essential services that we humans need from the environment of our planet, Earth. They are

- Unpolluted, fresh air
- Safe, fresh, drinking water
- Safe, nourishing food
- Renewable energy
- Waste recycling (any waste that cannot be recycled will accumulate in the environment)

2 Emergency

When a trauma victim is wheeled into the Emergency Room there needs to be a rapid assessment of the patient's status. This will involve establishing their vital signs and, in the most critical cases, transfer to intensive care.

Our planet, EARTH, is in the Emergency Room. Let's take a look at her vital signs.

O_2 Production

Approximately half of oxygen production occurs on land via trees and plants, and the other half in the oceans via phytoplankton. So important was the Amazon Rainforest for producing our oxygen that it used to be called 'the lungs of the Earth'. But, according to WWF, we are losing tropical rainforests at the astonishing rate of 30 soccer fields a minute!

Warming oceans over the past fifty years have led to a staggering 40% loss of phytoplankton, and inevitable further loss as global temperatures continue to rise.

Why haven't alarm bells been ringing long before now? If they have, why was nobody taking any notice?

Freshwater

Approximately 70% of our planet, Earth, is covered by water. So there should be no risk of water shortages. However, only 3% of that water is freshwater, and of that about two thirds is locked up in ice (WWF). Consequently, 2 billion people live in countries experiencing high levels of water stress. (UN 2019). Over half of the global population, 4.2 billion people, lack safely managed sanitation services, resulting in disease and death. (WHO/UNICEF 2019). At the current consumption rate two thirds of the world's population could be experiencing water shortages as early as 2025 (WWF).

Food Production

According to the WHO Global Nutrition Report (2018) every country in the world is affected by malnutrition. One-in-four people globally, that's 1.9 billion people, are moderately or severely food insecure (Roser and Ritchie, 2013).

We are unable to produce sufficient calories to sustain the existing global population. But the global population is expected to increase to about 10 billion by 2050 (UN Population Division), and so they say we need to double our calorie production to feed us all. However, we are already using all available arable land for crop production.

According to Dr Claire Kremen, Conservation Biologist, University of California, about 36% of crop production is used to feed animals, but only about 4% is returned to our food system in meat that we eat. We're losing about 90% of the calories going from grains to meat. And this is compounded by increasing demand for meat as economies develop. The single most important driver of increased meat consumption is increase in income per person (Milford and Le Mouël et al, 2019).

To quote Philip Wollen, former Vice-President of Citibank, "as I travel around I see poor countries who sell their grain to the West while their own children starve in their arms, and the

West feeds it to livestock so we can eat a steak. Am I the only one who sees this as a crime? Believe me every morsel of meat we eat is slapping the tear stained face of a hungry child."

In addition 63% of Amazon deforestation is for pasture for meat production.

Maybe we can look to fish to sustain us?

According to the World Bank (Sustainable Goals Atlas 2017) "Almost 90 percent of global marine fish stocks are now fully exploited or overfished."

It is clear that our patient's, Earth's, primary vital signs of oxygen, freshwater and food production are screaming at us that it is in need of urgent critical care.

But that is only part of the picture. Before embarking on treatment we need to also take a look at three important factors contributing to the problems we see in the primary vital signs, namely Climate Change, Biodiversity Loss and Pollution.

Climate Change

The evidence clearly shows that our patient, Earth, has been experiencing a rapid rise in global temperature and that this is set to continue, without urgent action. And the cause of this rapid rise is well understood. As can be seen by the accompanying graph of atmospheric CO_2, there has been a rapid rise from a pre-industrial level of 280ppm, to 416ppm as of February 7 2021 (https://www.co2.earth/daily-co2). In addition, air pollution from burning fossil fuels accounts for 8.7 million premature deaths a year (Vaughan, 2021).

If that wasn't bad enough atmospheric methane levels have also experienced a rapid rise, and its effect on global warming is 20x that of CO_2.

Furthermore, we now have a fully developed feedback loop. Vast quantities of methane had been locked up for millennia beneath the permafrost and ice. But now rising global temperatures have caused rapid thawing, thus releasing the locked up methane into the atmosphere, causing further warming and further release of methane.

"Climate change has both direct and indirect effects on agricultural productivity including changing rainfall patterns, drought, flooding and the geographical redistribution of pests and diseases. The vast amounts of CO_2 absorbed by the oceans causes acidification, influencing the health of our oceans and those whose livelihoods and nutrition depend on them" (UNFAO).

A destabilisation of weather patterns has already been experienced including, the scale and number of extreme weather events, eg hurricanes, droughts leading to massive fires, and floods.

But, more importantly, increasing global temperature is having a devastating impact on our patient's primary vital signs, oxygen production, freshwater and food production. Warming and increasing ocean acidity is causing a reduction in oxygen producing, photosynthesising marine plants, in turn leading to reductions in fish and pH sensitive marine life.

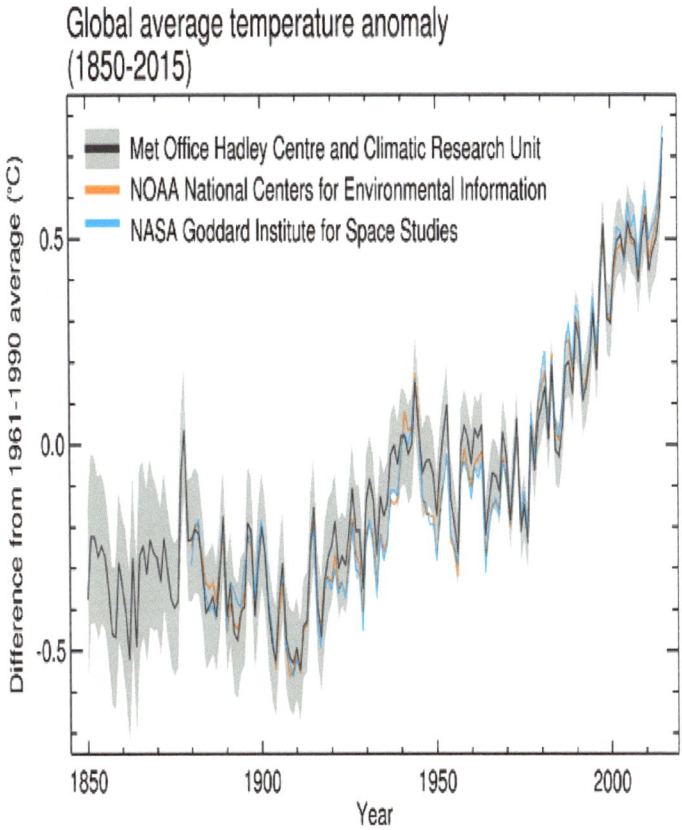

https://www.metoffice.gov.uk/weather/climate/science/global-temperature-records

Global CO_2

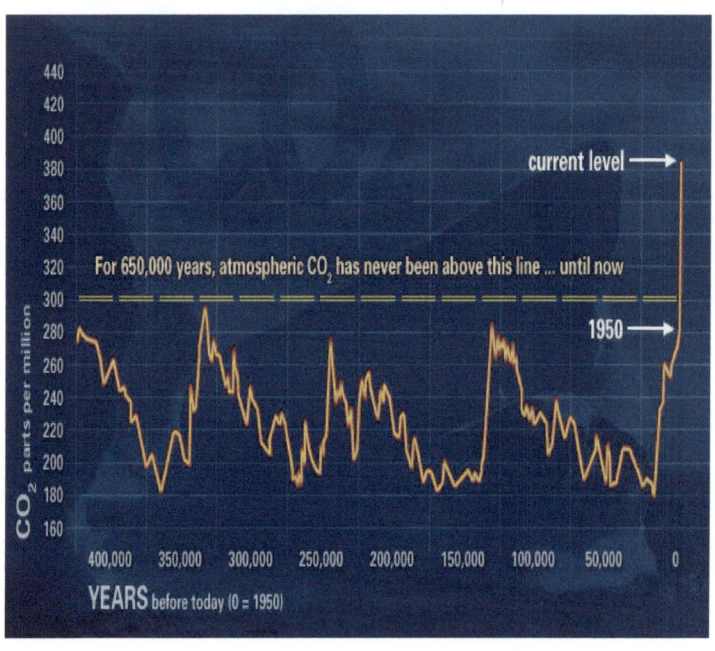

RECORD HIGH

Global emissions of methane have risen by nearly 10% over the past two decades, resulting in the highest-ever atmospheric concentrations of the greenhouse gas.

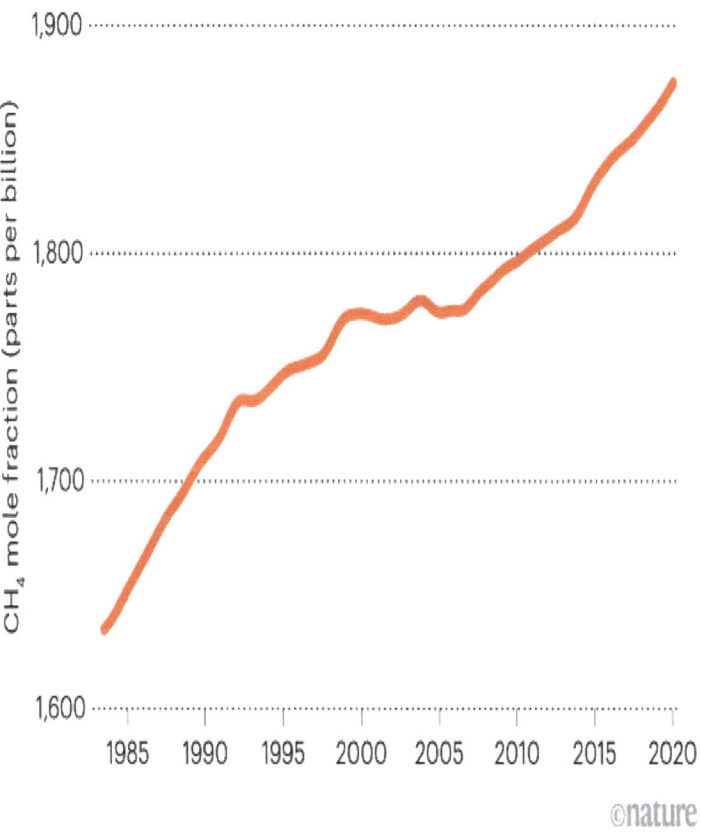

©nature

Biodiversity Loss

In the last 40 years we have lost more than half of the wildlife on the planet (WWF). We are experiencing a mass extinction that is progressing at an alarming pace.

This is not the first time there has been a mass extinction on our planet. However, the previous five mass extinctions were all due to natural causes, eg volcanic eruption, meteor strike. This is the first time in our planet's history that a single living species is on track to cause the extinction of almost all other species.

Why is that a problem?

We humans depend on biodiversity for oxygen, freshwater and food production. Many will already be aware of the alarm bells ringing at the loss of bee populations, so important for pollinating our food crops. Less well known is the intricate web of species needed to produce some foods.

As an example, consider the humble brazil nut. Brazil nut trees are found in undisturbed forest where orchids grow. The orchids are important as they are visited by the male orchid bee in order to obtain the scent he needs to attract the female orchid bee. In the process he pollinates the orchid. Only the female orchid bee has the body design that enables her to pollinate the brazil nut tree flower. During January and February the fruit, the size of a baseball, weighing 5lbs and containing 10 to 21 nuts, falls to the ground. The trees stand 160 feet tall and the fruit reaches speeds of up to 50 mph when it hits the ground, so the outer casing is extremely tough. So tough, that only the Agouti, a large rodent with just the right kind of teeth, can open it, which it does once it has carried it away and so enabled seed dispersal.

It is clear how delicate and vulnerable our interconnected ecosystems are to disruption. Yet we blindly and savagely destroy whole habitats in the reckless pursuit of economic growth.

Biodiversity also provides a measure of resilience in the face of challenges to our food supply by pests, disease and climate change. Clearly, the more species there are the more likely there will be one that can provide a solution to the impending threat.

But our growing population and increasing meat consumption continues to cause devastating habitat loss and hence loss of biodiversity. Livestock systems occupy 45% of Earth's surface and are a major driver of deforestation and habitat loss (UN).

Pollution

"Pollution is the largest environmental cause of disease and premature death. Pollution of air, land, and water cause more than 9 million premature deaths (16% of all deaths worldwide). That's three times more deaths than from AIDS, tuberculosis, and malaria combined and 15 times more than from all wars and other forms of violence." (World Bank) https://www.worldbank.org/en/topic/pollution

For far too long we have treated the planet that gives us life as a communal rubbish tip. Hazardous wastes from mining and industrial processes continue to pollute our air, land and water. Thanks to Sir David Attenborough's 'Blue Planet', global awareness of the problem has been raised, especially with regard to plastic pollution. There has been, and continues to be, progress to eliminate disposable and non-recyclable plastics.

Some governments have become involved in promoting recycling of materials. This not only reduces dumping of waste but also means less mining of raw materials, and its impact on habitat loss and pollution.

Summary

To summarize, we humans require our planet, Earth, to provide five services essential for our continuing survival. These are

- Fresh, unpolluted air.
- Fresh, unpolluted drinking water.
- Healthy, uncontaminated food.
- Energy, sufficient for lighting, heating, transport and to power our increasingly technology based life-style.
- Waste recycling.

All of which need to be provided in a sustainable way.

3 The Problem

Prior to the Covid19 pandemic, Climate Change was becoming a widely discussed problem, commanding media attention thanks in large part to Greta Thunberg. Greta became the poster child for Climate Change. Biodiversity Loss received much less attention. But Climate Change, Biodiversity Loss and Pollution are not the problem. They are all symptoms. The problem is consumption.

It is consumption that is driving all of the symptoms of our planet's destruction. And the problem of consumption is compounded by our growing population. Consumption, itself, is driven at the grassroots level by advertising, and at a national and international level by the narrow measure of economic development, GDP.

GDP, Gross Domestic Product, is defined as the final value of the goods and services produced within the geographic boundaries of a country during a specified period of time, normally a year.

The problem with GDP as a measure of economic development is that it values the environment, including habitats, as $zero. So, according to GDP, the Amazon rainforest has zero value until we cut down its trees and sell the timber, then convert the cleared space into pasture for raising cattle.

Clearly, the narrow measure of GDP leads to unsustainable development. That economic growth, coupled with human population, would exceed the physical limits of our planet was signalled way back in 1972 by Donella and Dennis Meadows and others in 'The Limits to Growth'. This gave birth to the concept of sustainable development, which was first incorporated into Bhutan's five year plan (1997-2002) as the four pillars.

- Sustainable and equitable economic development.
- Environmental conservation.
- Preservation and promotion of culture and heritage.
- Good governance.

There have been many attempts to incorporate environmental conservation into national and international measures of development, including 'Green GDP'. However, none has yet been adopted by governments for international comparison. Very recently, the UK Treasury commissioned a review into the economics of Biodiversity, in an attempt to begin to address the problem of the narrow definition of GDP (Dasgupta Review, 2021).

Whilst GDP remains the measure of development, unsustainable development will likely continue.

Some might hope that technological advancement will save us from our own destruction. Sadly, long before the required advancements have been developed and deployed at scale, we will have destroyed our planet.

The Covid19 pandemic provides some important lessons for us. The SARS-Cov-2 virus naturally spreads exponentially. Consequently, delays in acting to contain it have exponential consequences; exponential infections, exponential hospital admissions, exponential deaths, and exponential extensions to lockdowns to bring it back under control. Countries following this path rely on high-tech, vaccination solutions.

New Zealand acted swiftly and decisively, relying only on a low-tech lockdown, together with secure quarantine for all international arrivals. That has proved to be the most effective solution. New Zealand successfully eliminated the virus altogether and has had only 25 deaths, and the least time in lockdown.

Neither can we wait for governments to act to implement sustainable development, nor should we. It is, after all, our collective, individual consumption that is driving the destruction of our planet. Whilst others wait for high-tech solutions we can, like New Zealand, act swiftly and decisively to implement a low-tech solution. We, as citizens and custodians of our planet, must act, in our own best interests, to significantly reduce our consumption to sustainable levels.

This will have huge structural and economic consequences, resulting in significant job losses and a likely backlash from the denyers. Whole industries could be decimated. For example, consumption is fuelled by advertising.

Obviously, it makes sense to stop pouring fuel on to a fire. But ubiquitous advertising revenue supports almost all media. Advertising to children is illegal in some countries and strictly regulated in others for being unethical. But isn't it also unethical to stoke demand for consumption that is driving the destruction of the only planet capable of sustaining us?

Reducing consumption will impact all stages of the supply chain, eg mining, manufacturing, distribution and retailing. Travel and tourism, already severely impacted by the Covid19 pandemic, would need to continue to be restricted to essential journeys only. Unless and until sustainable travel can be developed and deployed at scale.

4 Needs vs Wants

Jeremy Brooks' 2013 article, 'Avoiding the limits to growth', begins with an account of his visit to a primary school in Bhutan, where there was a poster on the wall divided into two columns headed 'Needs' and 'Wants'. Under 'Needs' the children had listed food, water, oxygen, fire, shelter, clothing, and shoes. Remarkable, given that the school had no electricity. Under 'Wants' they had listed car, television, water boiler, new school dress, expensive pen, and gold.

Since the environmental emergency is being driven by consumption that far exceeds our planet's ability to supply, making a clear distinction between needs and wants is a most important first step. We need to be strict but realistic in this exercise. What was once a want, with progress in living standards, has become a need. For example, it would be unrealistic to expect those in developed countries to revert to washing their clothes in a nearby river. However, a car might be either a want or a need, depending on the specific circumstances.

This is my first attempt at a list of my needs.

Clean air to breath
Food - Adequate nutrition
Food storage facilities, ambient and cool
Drink
Maintain body temperature - clothes, heating, cooling
Water - wash body and laundry
Washing facilities inc wash kit with toothbrush+paste, towel
Laundry drying facility
Foot wear
Sanitation
Shelter / Security
Sleep facility
Sitting facility
Cooking and eating facilities
Lighting
Waste recycling / disposal
Hair / Beard - grooming / shaving
Hand + Toe nail clipping
Vision correction eg spectacles
Access to healthcare
Opportunity for exercise

Wet weather gear
Room Cleaning facilities eg broom, mop, dustpan+brush, duster
Access to communication facility, eg postbox, telephone
Access to News, eg radio
Access to transport
(Children need access to education)
Access to social interaction

The house I moved into five years ago has a built-in fan assisted oven with grill and electric hob, together with a built-in dishwasher, neither of which I have used. Most of my cooking is by more fuel-efficient microwave, and I hand-wash my dishes.

5 *Think!* EARTH

We need to adopt the habit to '*Think!* EARTH' in every decision and choice that we make, so as to make choices that minimise our impact on the only planet capable of sustaining us.

The decision that will cause the greatest damage to the planet, that far exceeds all others combined, is to bring a child into the world (Wynes and Nicholas 2017).

Obviously, in most circumstances, buying used will be less destructive for our planet than buying new. Even if the item has already been made, our purchase at some point will trigger stock replenishment and hence manufacture. When we buy new we incur all of the detrimental effects of material consumption. This might include destruction of habitat for mining, energy transporting raw material for processing, energy and water for processing, dumping of waste from processing, transport to manufacture, more energy, water and waste during manufacture, transport for distribution etc. Of course, if manufacturers use exclusively recycled material and exclusively renewable energy, then the damage to our planet will be significantly reduced.

Now let us consider some specific decisions and choices with their built-in damage to our planet.

Deciding where to live

If we choose to live in the USA where the average CO_2 emissions per person per annum are 16 tons (Nature Conservancy), we can expect that our damage to our planet will be significantly greater than if we choose to live in Chad where the average CO_2 emissions per person per annum are 0.06 tons (Ibrahim 2020).

More developed countries place a far greater burden on our planet than lesser developed countries. As lesser developed countries strive to 'progress' the damage they cause to our planet increases, unless such progress is achieved within the constraints of sustainability.

At a more mundane level, choosing to live near our place of work, such that we can walk or cycle to work, will obviously limit the energy needed for the daily commute.

Choosing Accommodation

In a like for like comparison, a new build will cause more damage to our planet than an existing property. However, a property that supports living with a zero carbon footprint is kinder to our planet. So, if comparing a new build zero carbon property with an existing energy inefficient property, we might need to consider the life cycle environmental impact of the new build against the environmental impact of the remaining lifetime of the existing property. But before we do we should take account of the opportunities to improve the energy efficiency of the existing property, eg via insulation.

For existing housing stock, again on an energy efficient like for like comparison, the damage caused can be ranked in ascending order of apartment, mid-terraced house, end-terraced house, semi-detached house, detached house.

Choosing Home Energy

We should choose renewable energy whenever that affordable option is available to us.

Deciding what to eat and drink

As we saw when reviewing the state of the emergency facing us, consuming land or water-based animals and animal products, eg milk, is unsustainable for the size of our global population. We need to choose plant-based foods and drink.

Deciding what to wear

But consumption of animals extends far beyond food. The cosmetics industry has a long history of using animal-based products, eg lipsticks from crushed red cochineal insects and perfume from ambergris (whale secretions), in addition to testing products on animals. Fashion has moved away from adorning us with animal furs and skins. We no longer need to wear mink eyelashes, and there is also the option for leather made from plant materials or recycled plastic bottles.

Travel

A simple checklist can make a significant difference to the damage we do to our planet as a consequence of our travel decisions.

For every journey we need to consider,

Is the journey essential?

Is walking or cycling an option?

Is public transport an option?

And if we must have our own vehicle, which one has the least life cycle environmental impact?

For longer journeys the choices in ascending order of impact to the environment are
Train
Plane
Ship

(I was surprised to find that ships are more environmentally damaging than planes.)

Consumer Products and Services

Our first step should be to check our list of needs. If it's not on the list, then we should check the total, eg life cycle, damage our planet suffers as a consequence of this product or service, and let that determine our decision, should we choose to proceed with the purchase.

Education, Training and Working

Our choice of education and training will have a significant effect on the course our life takes and, as a consequence, the damage that we will do to our planet. We could choose a course that involves working to improve the environment or ecology. We could choose a trade or profession and then work in a context that serves to improve the environment directly or indirectly. (But see also the section on Charities.) It would be best to avoid education, training and working for organizations directly or indirectly involved in damaging our planet, eg fossil fuel extraction or advertising.

Pets

For most of us a pet would be classed as a want rather than a need. However, there are specific circumstances where a pet might be considered a need. Any pet will have life cycle consequences for our planet, eg the production of their 'home' and their food.

There could also be additional consequences. For example, domestic cats were estimated to account for the loss of over one billion birds and over six billion small mammals each year in the USA alone (Loss et al, 2013). More recent research suggests that feeding cats meat reduces their wild predation, but not without its own consequences for our planet (Cecchetti et al, 2021).

Charities

In recent years there has been a growing movement towards 'Effective Altruism', which basically means being generous with our time and money to do the most good. As an example of this, I read about a man who wanted to save lives and initially thought of studying to become a doctor. After some research he discovered that he could save more lives by donating £600 pa to a charity providing mosquito nets.

If we are able to donate to charities it would be best to choose those most likely to have the greatest impact on improving our planet's environment. Hitherto, my priority has been to alleviate human suffering, eg displaced refugees, the homeless, the hungry and the sick. I need to reflect on the greatest overall relief to mankind's suffering as a consequence of shorter term goals or longer term goals, or maybe a balance of the two.

Useful advice on 'Which Career' and 'Which Charity' can be found at https://www.effectivealtruism.org/articles/introduction-to-effective-altruism/

6 Conclusion

We have seen that it is as though our planet has just been wheeled into the Emergency Room and will suffer premature death unless urgent life-saving action is taken. The lives of each and every one of us hang in the balance, and critically depend on what each and every one of us now does. With concerted, some might say sacrificial, effort we can save our planet, for ourselves and for future generations.

It is clear that it is consumption compounded by population that is the key driver for our planet's destruction. Consequently, we must drastically reduce our consumption and procreation to a sustainable level. Effectively, this means a clear distinction between needs and wants, and consuming only that which is necessary to meet our needs.

To do so means adopting the habit to *Think*! EARTH in every decision and choice that we make.

Bibliography

Allen, E. (2021) Veganism is not just for January, and it's not just about food, Harper's Bazaar, Feb 1, 2021. https://www.harpersbazaar.com/uk/fashion/a35380171/veganism-is-not-just-about-food/

Brooks, J. S. (2013) Avoiding the Limits to Growth: Gross National Happiness in Bhutan as a Model for Sustainable Development, Sustainability 2013, 5, 3640-3664; doi:10.3390/su5093640

Cecchetti, M., Crowley, S. L., Goodwin, C. E. D., McDonald, R. A. (2021) Provision of High Meat Content Food and Object Play Reduce Predation of Wild Animals by Domestic Cats Felis catus. Current Biology, 2021; DOI: 10.1016/j.cub.2020.12.044

Dasgupta, P. (2021) Final Report - The Economics of Biodiversity: The Dasgupta Review. HM Treasury, 2 February 2021.

Ibrahim, H. O. (2020) in Royal Institution Christmas Lectures 2020 - Planet Earth - A User's Guide, 3 Up in the Air, 30 December 2020, presented by Dr Tara Shine.

Loss, S. R., Will, T., Marra, P. P. (2013) The impact of free-ranging domestic cats on wildlife in the United States. Nature Communications 4, article no 1396(2013).

Meadows, D. H., Meadows, D.L., Randers, J., Behrens III, W. W. (1972) Limits to Growth. Chelsea Green Publishing Co.: White River Junction, VT, USA.

Milford, A. B., Le Mouël, C., Bodirsky, L. B., Rolinski, S. (2019) Drivers of meat consumption, Appetite, Vol 141, 1 October 2019, 104313 https://www.sciencedirect.com/science/article/abs/pii/S0195666319301047

Planetary Emergency Plan: Securing a New deal for People, Nature and Climate, the Club of Rome in partnership with the Potsdam Institute for Climate Impact research. The Club of Rome Lagerhausstrasse 9 CH-8400 Winterthur SWITZERLAND.

https://clubofrome.org/wp-content/uploads/2020/02/PlanetaryEmergencyPlan_CoR-4.pdf

Roser, M. and Ritchie, H. (2013) Hunger and Undernourishment. Online at OurWorldInData.org, retrieved from 'https://ourworldindata.org/hunger-and-undernourishment'

Vaughan, A. (2021) Deaths from fossil fuel air pollution are double what we thought. New Scientist 9 February 2021. https://www.newscientist.com/article/2267035-deaths-from-fossil-fuel-air-pollution-are-double-what-we-thought/?utm_source=nsday&utm_medium=email&utm_campaign=NSDAY_100221

Watts, S. (2018) Climate Change is putting the ocean's phytoplankton in danger, This Week Jan 10, 2018.

World Bank Atlas of Sustainable Development Goals 2017, Life below water. https://datatopics.worldbank.org/sdgatlas/archive/2017/SDG-14-life-below-water.html

Wynes, S. and Nicholas, K. A. (2017) The climate mitigation gap: education and government recommendations miss the most effective individual actions, Environmental Research Letters 12 July 2017 • © 2017 IOP Publishing Ltd,, Volume 12, Number 7

www.ingramcontent.com/pod-product-compliance
Lightning Source LLC
Chambersburg PA
CBHW040252220526
45473CB00001B/453